INTRODUCTION

North American coastlines are home to hundreds of species of resident and migrant saltwater fishes. Of these, a select few are the target of anglers and skippers seeking the ultimate battle with some of nature's largest and most beautiful game fish. From the nearshore species like striped bass, tarpon, salmon and bluefish, to the giant offshore exotics including marlin, sailfish, tuna and swordfish, this guide provides fish enthusiasts of all levels with a simplified, pocket-sized reference to some of the most popular and sought-after saltwater game fish found in our coastal waters.

Before Heading Out

To enjoy the wealth of fishing opportunities from your local pier or beach to offshore hotspots, you only need to follow a few simple rules:

1. Purchase a fishing license if needed (regulations vary by state). Also purchase necessary tags to take specific species if needed.
2. Ensure it is legal to fish your location and that you are using legal baits and approved fishing procedures.
3. Be aware of minimum size and bag limits for each species you keep.

PARTS OF A TUNA

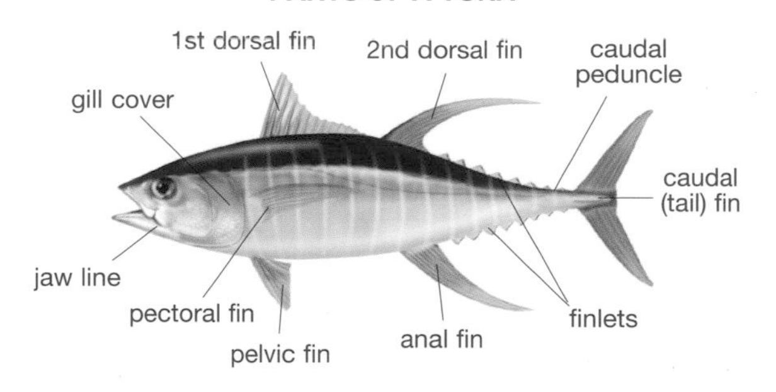

Most illustrations show the adult male in breeding coloration. Colors and markings may be duller or absent during different seasons. The measurements denote the approximate maximum length of species. Illustrations are not to scale.

Made in the USA

Waterford Press publishes reference guides that introduce readers to nature observation, outdoor recreation and survival skills. Product information is featured on the website: **www.waterfordpress.com**

To order or for information on custom published products please call 800-434-2555 or email orderdesk@waterfordpress.com. For permissions or to share comments email editor@waterfordpress.com.

ISBN 978-1-62005-419-2
$7.95 U.S.
50795
9 781620 054192
UPC
8 84682 01461 2
10 9 8 7 6 5 4 3 2 1
213201

SALTWATER GAME FISH – An Illustrated Folding Pocket Guide to Familiar Species

Kavanagh/Leung

SALTWATER GAME FISH

of North America

An Illustrated Folding Pocket Guide to Familiar Species

Striped Bass *Morone saxatilis*

Description: Silvery-sided fish has 6-9 dark, uninterrupted side stripes. **Length**: To 6 ft. (1.8 m) **Weight**: To 82 lbs. (37 kg) **Habitat**: Nearshore coastal waters, moves into freshwater to spawn. **Range:** Native to the east coast, it has been widely introduced throughout North America. **Comments**: Introduced to the Pacific coast in the 1880s. An important sport and food fish.

Sailfish *Istiophorus platypterus*

Description: Told at a glance by its huge, fan-shaped, retractable dorsal fin. Has long and pointed rigid snout that forms a bill or spear. **Length**: To 11 ft. (3.3 m) **Weight**: To 220 lbs. (100 kg) **Habitat**: Warm oceanic waters. **Range**: From RI to Brazil, San Diego to Chile. **Comments**: A predator, it uses its bill to slash at or stab fish.

Blue Marlin *Makaira nigricans*

Description: Dark blue above and lighter below, it has a rigid dorsal fin and about 15 bars on its sides. Note steep head profile. Has long and pointed rigid bill. **Length**: To 14 ft. (4.2 m) **Weight**: To 1,400 lbs. (636 kg) **Habitat**: At surface to middle depths in open oceans. **Range**: From ME to Gulf of Mexico to Uruguay, rare in west from southern CA to Chile. **Comments**: A prized game fish renowned for its fighting ability.

Swordfish *Xiphias gladius*

Description: Elongate, rounded fish has an extremely long and pointed rigid snout/sword and a short-based, non-retractable dorsal fin. **Length**: To 15 ft. (4.5 m) **Weight**: To 1,182 lbs. (536 kg) **Habitat**: Surface waters to depths of 3,000 ft. (900 m). **Range**: In Atlantic from NL to Argentina; Pacific OR to Chile. **Comments**: Feeds on cephalopods and fishes. A highly prized game fish that is very difficult and expensive to pursue.

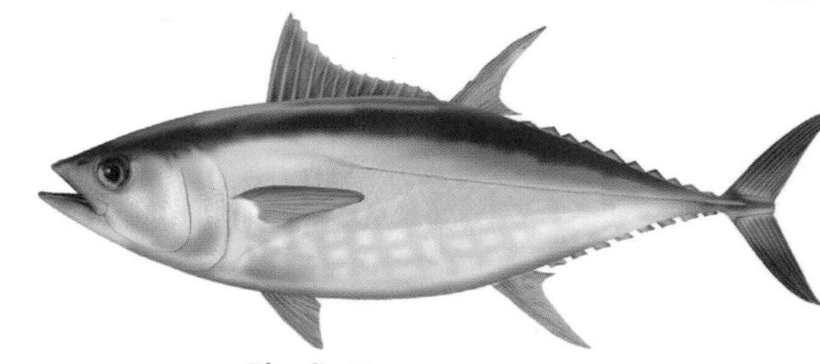

Bluefin Tuna *Thunnus thynnus*

Description: Robust, torpedo-shaped fish is blue-black above. Finlets following dorsal and anal fins are yellowish. **Length**: To 10 ft. (3 m) **Weight**: To 1500 lbs. (679 kg) **Habitat**: Surface of open oceans. **Range:** In Atlantic from Labrador to Brazil including Gulf of Mexico; Pacific from AK to Peru. **Comments**: A highly prized food fish in Asia, a single fish sold for $1.8 million in 2013 in Japan.

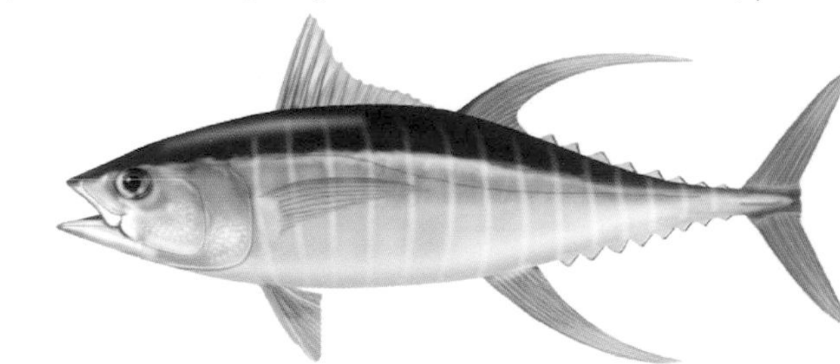

Yellowfin Tuna *Thunnus albacares*

Description: Robust, torpedo-shaped fish. Second dorsal fin and finlets following dorsal and anal fin are yellow. Note yellow side stripe. **Length**: To 7 ft. (2.1 m) **Weight**: To 427 lbs. (194 kg) **Habitat**: Surface and mid-depths of oceans. **Range**: In Atlantic from MA to Brazil; Pacific from CA to Chile. **Comments**: Perhaps the most valuable commercial species, it is also a prized sport fish.

Bonito *Sarda* spp.

Description: Bluish above and silvery below, it has oblique side stripes and 7-9 finlets following dorsal and anal fins. **Length**: To 40 in. (1 m) **Weight**: To 21 lbs. (.7 kg) **Habitat**: Near surface in tropical and temperate oceans. **Range**: In Atlantic from NS to Argentina including Gulf of Mexico; Pacific from AK to Chile. **Comments**: Four similar species live in Atlantic and Pacific oceans.

Albacore *Thunnus alalunga*

Description: Note pointed snout and long pectoral fin. **Length**: To 5 ft. (1.5 m) **Weight**: To 88 lbs. (40 kg) **Habitat**: Open ocean at surface and mid-depths. **Range**: NS to Brazil, AK to Mexico. **Comments**: One of the most prized Pacific sport fish, it is common during the summer months.

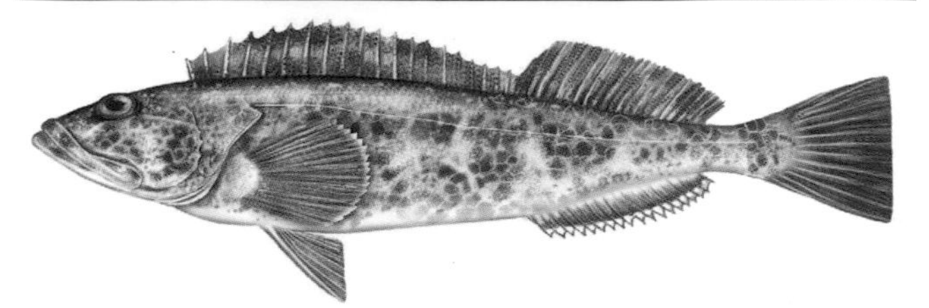

Lingcod *Ophiodon elongatus*

Description: Elongate, rounded fish is mottled gray-brown to green. Lower jaw projects beyond upper jaw. **Length**: To 5 ft. (1.5 m) **Weight**: To 82 lbs. (37 kg) **Habitat**: Over reefs and soft bottoms to depths of 1,400 ft. (427 m) **Range:** AK to Baja, CA. **Comments**: A voracious predator, it is considered one of the best food fish and is also important commercially.

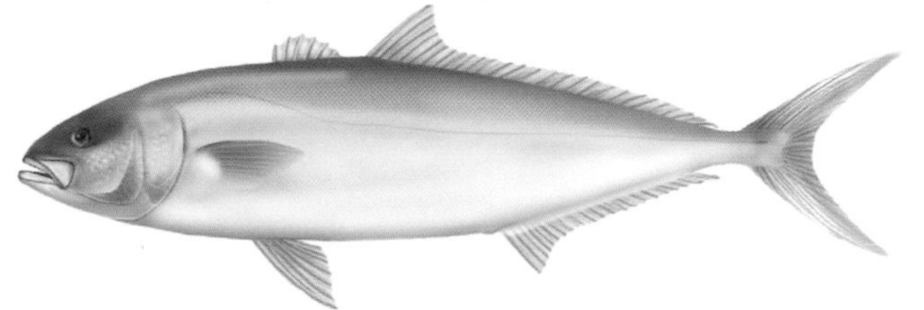

Yellowtail *Seriola lalandi*

Description: Large silvery fish has a yellow side stripe and yellow fins. Note short, spiny first dorsal fin. **Length**: To 5 ft. (1.5 m) **Weight**: To 114 lbs. (52 kg) **Habitat**: Surface waters near reefs, islands and kelp beds. **Range:** BC to Chile. **Comments**: Feeds on small schooling fish and is common in nearshore waters. Very popular game fish in southern California.

Kelp Bass *Paralabrax clathratus*

Description: Green-brown above, it has white blotches between the dorsal fin and lateral line. **Length**: To 30 in. (75 cm) **Weight**: To 14.5 lbs. (6.5 kg) **Habitat**: In or near kelp beds to depths of 150 ft. (46 m) **Range**: OR to Baja, CA. **Comments**: Also called calico bass, it is very common in its southern range where annual catches have exceeded one million.

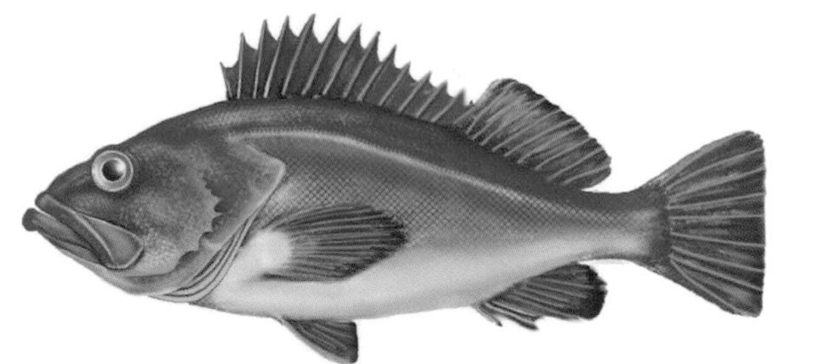

Yelloweye Rockfish *Sebastes ruberrimus*

Description: Orange fish has bright yellow eyes. Fin edges are black. Note light side stripe (younger individuals have two light side stripes). **Length**: To 3 ft. (90 cm) **Weight**: To 39 lbs. (18 kg) **Habitat**: Nearshore waters to depths of 2,000 ft. (600 m). **Range**: AK to Baja, CA. **Comments**: Highly prized sport and food fish.

Coho Salmon *Oncorhynchus kisutch*

Description: Has dark spots on its back and the upper lobe of its caudal fin. Gums are white at tooth base. Breeding male has red side stripes. **Length**: To 40 in. (1 m) **Weight**: To 36 lbs. (16 kg) **Habitat**: Shallow to mid-level ocean waters. Returns to coastal streams to spawn, often far inland. **Range**: Bering Strait to Baja. Widely introduced. **Comments**: Also called silver salmon.

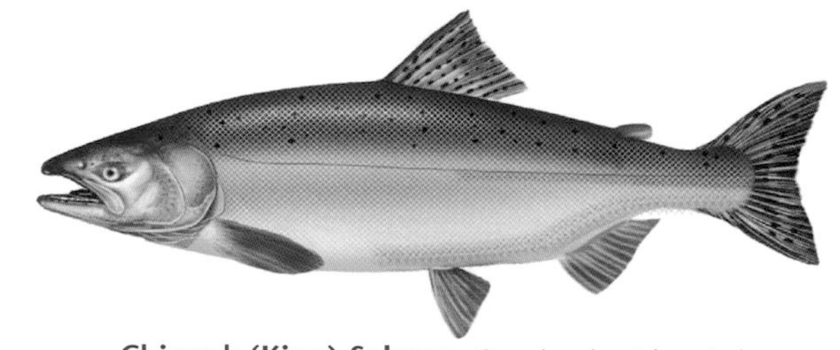

Chinook (King) Salmon *Oncorhynchus tshawytscha*

Description: Has dark spots on its back, dorsal and adipose fins and caudal fin. Gums are black at tooth base. **Length**: To 5 ft. (1.5 m) **Weight**: To 100 lbs. (45 kg) **Habitat**: Shallow to mid-level ocean waters. Returns to large rivers to spawn. **Range:** Bering Strait to southern CA. Widely introduced. **Comments**: The largest salmon, it is also called king salmon.

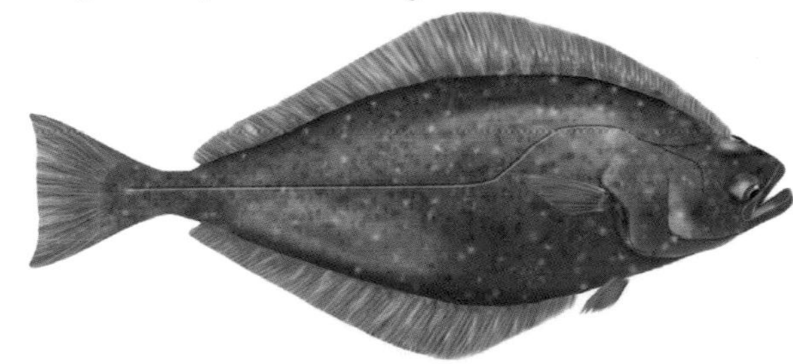

Pacific Halibut *Hippoglossus stenolepis*

Description: Huge mottled-brown flatfish. A right-eyed flounder, its eye is usually on the right side of the body. Underside is white. **Length**: To 9 ft. (2.7 m) **Weight**: To 459 lbs. (208 kg) **Habitat**: Bottoms to depths of 3,600 ft. (1097 m) **Range:** AK to CA. **Comments**: A prized sport, food and commercial fish. The smaller California halibut (*Paralichthys californicus*) is a left-eyed flounder.

Roosterfish *Nematistius pectoralis*

Description: Easily distinguished by long, threadlike spines at front of dorsal fin and two curved side stripes. **Length**: To 4 ft. (1.2 m) **Weight**: To 114 lbs. (52 kg) **Habitat**: Shallow, nearshore waters. **Range:** Southern CA to Peru, including Galapagos Islands. **Comments**: A prized sport fish and food fish.

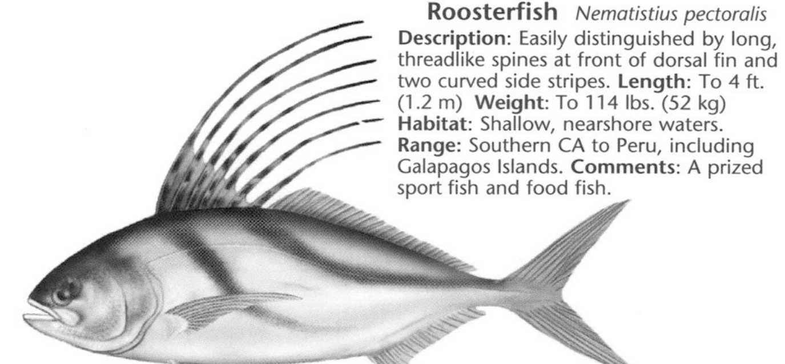

Red Drum *Sciaenops ocellatus*

Description: Reddish to copper colored fish has a dark spot at base of tail fin. **Length:** To 5 ft. (1.5 m) **Weight:** To 94 lbs. (46 kg) **Habitat:** Surf zone to offshore waters. **Range:** MA to Mexico including Gulf of Mexico. **Comments:** A prized game fish for surf-casters. Also known as redfish, channel bass or simply reds.

Spotted Seatrout *Cynoscion nebulosus*

Description: Blue-gray above and silvery-white below, it is black-spotted on its upper side, including dorsal and caudal fins. **Length:** To 28 in. (70 cm) **Weight:** To 17 lbs. (8 kg) **Habitat:** Estuaries, marshes, shallow coastal waters. **Range:** MA to FL including Gulf of Mexico. **Comments:** Also called speck, it is an important food and game fish.

Weakfish *Cynoscion regalis*

Description: Dark blue to olive above, it has rows of small spots that form oblique lines on upper body. Lower jaw projects beyond upper jaw. **Length:** To 3 ft. (90 cm) **Weight:** To 19 lbs. (9 kg) **Habitat:** Shallow coastal waters, estuaries. **Range:** VA to FL. **Comments:** Also called summer trout and tiderunner, it is an important food, game and commercial fish that is especially abundant from NC to FL.

Bluefish *Pomatomus saltatrix*

Description: Green to blue above and silvery below, it has a large head and mouth with prominent, sharp teeth. Note dark blotch at base of pectoral fin. **Length:** To 43 in. (1.09 m) **Weight:** To 32 lbs. (14 kg) **Habitat:** Nearshore and offshore waters near surface. **Range:** NS to Argentina including Gulf of Mexico. **Comments:** Voracious predators, they follow schools of small fish into nearshore waters and have been known to bite people when feeding.

Tripletail *Lobotes surinamensis*

Description: Tan to dark brown, deep-bodied fish has elongate dorsal and anal fins that give it a 'triple-tailed' appearance. **Length:** To 40 in. (1 m) **Weight:** To 42 lbs. (19 kg) **Habitat:** Nearshore waters, estuaries. **Range:** MA to Argentina including Gulf of Mexico. **Comments:** Often seen floating at the surface on its side. An esteemed food fish.

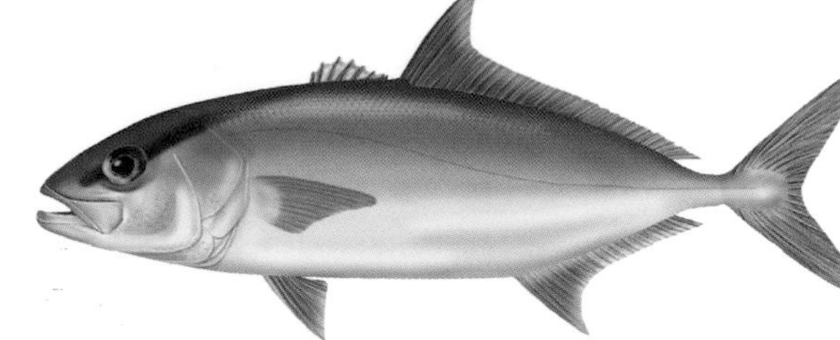

Greater Amberjack *Seriola dumerili*

Description: Note dark stripe from snout through eye toward dorsal fin. **Length:** To 6 ft. (1.8 m) **Weight:** To 163 lbs. (74 kg) **Habitat:** Coastal oceans particularly around wrecks. **Range:** MA to Brazil including Gulf of Mexico. **Comments:** An important commercial and sport fish.

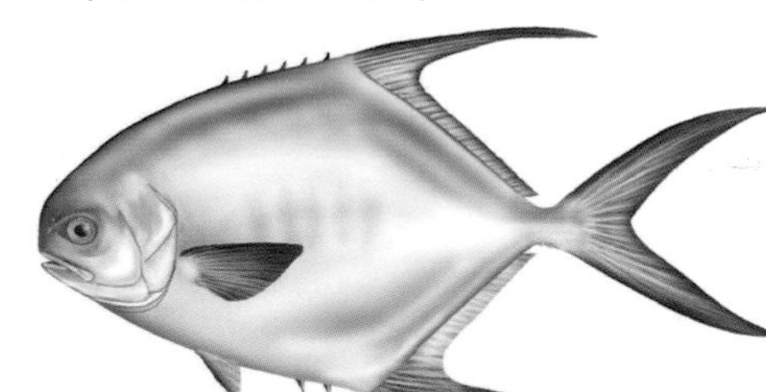

Permit *Trachinotus falcatus*

Description: Deep-bodied, silvery fish has a very low, spiny first dorsal fin; second dorsal fin has a single spine. **Length:** To 39 in. (98 cm) **Weight:** To 60 lbs. (27 kg) **Habitat:** Shallow water to depths of 100 ft. (30 m). **Range:** MA to Brazil including Gulf of Mexico. **Comments:** Very similar to the Florida pompano (*T. carolinus*) which is smaller and has a yellowish belly. Found in the same waters.

Crevalle Jack *Caranx hippos*

Description: Deep-bodied greenish-blue fish has a yellowish belly. Note dark spot on gill cover and pectoral fin. Head has steep profile. **Length:** To 40 in. (1 m) **Weight:** To 66 lbs. (30 kg) **Habitat:** Nearshore and offshore waters. **Range:** NS to Uruguay including Gulf of Mexico. **Comments:** Also called common jack, it enters freshwater in Florida.

Tarpon *Megalops atlanticus*

Description: Huge, silvery, large-headed fish has a jutting lower jaw and a large mouth. Silvery scales are extremely large. **Length:** To 8 ft. (2.4 m) **Weight:** To 286 lbs. (130 kg) **Habitat:** Nearshore and offshore coastal waters. **Range:** NS to Brazil, including Gulf of Mexico. **Comments:** The 'silverking' of the sport fish, it is a fast swimmer that makes spectacular leaps once hooked.

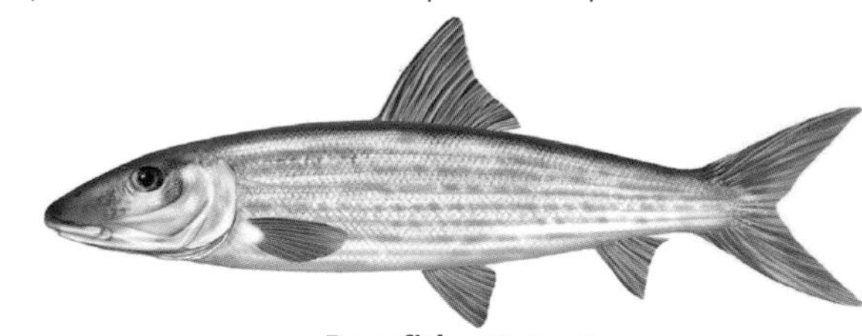

Bonefish *Albula vulpes*

Description: Torpedo-shaped, silvery fish. Snout overhangs lower jaw. Caudal fin is deeply forked. **Length:** To 3 ft. (90 cm) **Weight:** To 16 lbs. (7.3 kg) **Habitat:** Shallow waters over soft bottoms. **Range:** Northern Florida to Brazil including Gulf of Mexico. **Comments:** Most common in south FL, Bermuda and the Bahamas.

Atlantic Salmon *Salmo salar*

Description: Has black spots (often X-shaped) on sides (but not on fins) and 2-3 large spots on gill cover. Marine form is silvery. Freshwater form is brown-bronze and is reddish on back and side. **Length:** To 4.5 ft. (1.4 m) **Weight:** To 80 lbs. (36 kg) **Habitat:** Coastal waters, freshwater lakes and streams. **Range:** Native from Arctic Circle through N Quebec to Connecticut River. Many landlocked populations in New England. Extensively farmed on east and west coasts in Canada. **Comments:** Unlike Pacific salmon, it does not die after spawning and returns to the sea.

Snook *Centropomus undecimalis*

Description: Slender, elongate fish has a pointed snout and jutting lower jaw. Dark, sloping lateral line is prominent. **Length:** To 4 ft. (1.2 m) **Weight:** To 54 lbs. (24 kg) **Habitat:** Shallow coastal waters, enters freshwater occasionally. **Range:** NC to Brazil including Gulf of Mexico. **Comments:** Highly esteemed sport and food fish, feeds on other fish and crustaceans.

Atlantic Cod *Gadus morhua*

Description: Rounded green-brown to yellowish fleshy fish has a large chin barbel 'whisker'. Upper jaw projects beyond lower. Note three dorsal and two anal fins. **Length:** To 4 ft. (1.2 m) **Weight:** To 104 lbs. (47 kg) **Habitat:** Along bottoms from 40-1800 ft. (548 m) deep. **Range:** Greenland to NC. **Comments:** The similar Pacific cod (G. macrocephalus) is common on the west coast.

Cobia *Rachycentron canadum*

Description: Long, dark brown fish with a flattened head has a dark side stripe. Low, first dorsal fin has 7-9 spines. Caudal fin is deeply forked. **Length:** To 7 ft. (2.1 m) **Weight:** To 135 lbs. (61 kg) **Habitat:** Coastal and offshore waters, often around reefs and barrier islands. **Range:** MA to Argentina, including Gulf of Mexico. **Comments:** Also called lemonfish, ling, black salmon, black kingfish, crab eater, runner.

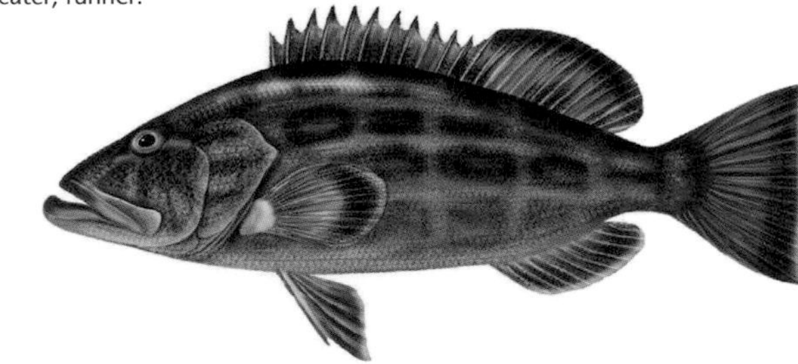

Black Grouper *Mycteroperca bonaci*

Description: Robust, wide-bodied fish is gray to dark brown with irregular rows of rectangular, darker blotches. Edges of caudal, anal and second dorsal fin are black. **Length:** To 4 ft. (1.2 m) **Weight:** To 124 lbs. (56 kg) **Habitat:** Rocky and coral reefs. **Range:** FL to Brazil. **Comments:** Also called marbled rockfish, it is an esteemed food fish.

Goliath Grouper *Epinephelus itajara*

Description: Huge, robust mottled fish has black spots, irregular side bands and a rounded caudal fin. **Length:** To 8 ft. (2.4 m) **Weight:** To 680 lbs. (308 kg) **Habitat:** Shallow water to depths of 100 ft. (30 m). **Range:** Primarily FL to Brazil, including Gulf of Mexico, occasionally caught off New England. **Comments:** Feeds on crustaceans, other fish and young turtles and is known to attack divers. A prized table fish.

Red Snapper *Lutjanus campechanus*

Description: Distinguished by pinkish color and sharply pointed anal fin. **Length:** To 3 ft. (90 cm) **Weight:** To 50 lbs. (23 kg) **Habitat:** Over rocks and reefs to depths of 600 ft. (180 m). **Range:** MA to Yucatan including Gulf of Mexico. **Comments:** One of America's most important commercial and sport fishes.

Great Barracuda *Sphyraena barracuda*

Description: Elongate, torpedo-shaped, silvery fish has a jutting lower jaw and exposed, fang-like teeth. **Length:** 6 ft. (1.8 m) **Weight:** To 87 lbs. (40 kg) **Habitat:** Inshore and offshore waters. **Range:** From MA to Brazil including the Gulf of Mexico. **Comments:** A voracious predator, it feeds primarily on other fishes. Is attracted to and will bite at shiny objects like jewelry. The similar Pacific barracuda (*S. argentea*) is one of the top game fish in southern California.

Dolphinfish *Coryphaena hippurus*

Description: Unmistakable snub-nosed fish is iridescent blue or green and has yellow sides flecked with spots. Note steep forehead and long dorsal and anal fins. **Length:** To 7 ft. (2.1 m) **Weight:** To 87 lbs. (40 kg) **Habitat:** Primarily open oceans over deep water. **Range:** NS to Brazil including Gulf of Mexico. **Comments:** Also known as mahi-mahi and dorado. Once landed, its iridescent colors fade quickly. Esteemed sport and food fish.

Wahoo *Acanthocybium solandri*

Description: Long, slender fish with a beak-like, toothy mouth is steely-blue to dark greenish above and has numerous dark side bars. The well-defined lateral line dips sharply near the middle of the first dorsal fin. **Length:** To 7 ft. (2.1 m) **Weight:** To 184 lbs. (83 kg) **Habitat:** Surface of offshore waters. **Range:** NY to Brazil including Gulf of Mexico. **Comments:** Esteemed for its fighting ability and excellent flavor. Also called kingfish.

Spanish Mackerel *Scomberomorus maculatus*

Description: Elongate fish is dark blue above and silvery below with yellowish side spots. First dorsal fin is black. **Length:** To 3 ft. (90 cm) **Weight:** To 13 lbs. (6 kg) **Habitat:** Near surface of open seas and nearshore waters. **Range:** ME to Yucatan. **Comments:** A very important commercial fish, it follows schools of bait fish which are often indicated by excited birds flying overhead.

King Mackerel (Kingfish) *Scomberomorus cavalla*

Description: Is dark bluish above and silvery below. First dorsal fin is silvery. Note sharp drop in lateral line under second dorsal fin. **Length:** To 5 ft. (1.5 m) **Weight:** To 93 lbs. (42 kg) **Habitat:** Warm, open oceans and nearshore waters. **Range:** ME to Florida and the Gulf of Mexico. **Comments:** The target of many fishing tournaments, it is an esteemed sport and food fish.

World Records on All Tackle

Fish	Weight	Metric	Location
Albacore	88 lbs. 2 oz.	(39.97 kg)	Canary Islands
Atlantic Cod	103 lbs. 10 oz.	(47.02 kg)	Norway
Atlantic Salmon	79 lbs. 2 oz.	(35.89 kg)	Norway
Black Grouper	124 lbs. 0 oz.	(56.24 kg)	TX, USA
Blue Marlin	1402 lbs. 2 oz.	(636 kg)	Brazil
Bluefin Tuna	1496 lbs. 0 oz.	(678.58 kg)	NS, Canada
Bluefish	31 lbs. 12 oz.	(14.4 kg)	NC, USA
Bonefish	16 lbs. 0 oz.	(7.26 kg)	Bahamas
Bonito	21 lbs. 5 oz.	(9.67 kg)	CA, USA
Chinook Salmon	97 lbs. 4 oz.	(44.11 kg)	AK, USA
Cobia	135 lbs. 9 oz.	(61.5 kg)	Australia
Coho Salmon	33 lbs. 4 oz.	(15.08 kg)	NY, USA
Dolphinfish	87 lbs. 0 oz.	(39.46 kg)	Costa Rica
Goliath Grouper	680 lbs. 0 oz.	(308.44 kg)	FL, USA
Greater Amberjack	163 lbs. 2 oz.	(74 kg)	Japan
King Mackerel	93 lbs. 0 oz.	(42.18 kg)	Puerto Rico
Pacific Halibut	459 lbs. 0 oz.	(208.2 kg)	AK, USA
Red Drum	94 lbs. 2 oz.	(42.69 kg)	NC, USA
Roosterfish	114 lbs. 0 oz.	(51.71 kg)	Baja, CA
Sailfish	221 lbs. 0 oz.	(100.24 kg)	Ecuador
Snook	53 lbs. 10 oz.	(24.32 kg)	Costa Rica
Spanish Mackerel	13 lbs. 0 oz.	(5.89 kg)	NC, USA
Spotted Seatrout	17 lbs. 7 oz.	(7.92 kg)	FL, USA
Striped Bass	81 lbs. 14 oz.	(37.14 kg)	CT, USA
Swordfish	1182 lbs. 0 oz.	(536.15 kg)	Chile
Tarpon	286 lbs. 9 oz.	(129.98 kg)	Guinea-Bissau
Wahoo	184 lbs. 0 oz.	(83.46 kg)	Baja, CA
Yellowfin Tuna	427 lbs. 0 oz.	(193.68 kg)	Baja, CA
Yellowtail	114 lbs. 10 oz.	(52 kg)	New Zealand

Source: International Game Fish Association (IGFA)